Introduction to Densitometry

A User's Guide to Print Production Measurement Using Densitometry

Peter V. Brehm
Vice President/Print Technologies
Graphic Communications Association

3rd Revision, May 1992

Library of Congress Catalog Card Number: 90-81119
ISBN 0-933505-10-8

Printed in the United States of America

Designed by Tom Semmes

Acknowledgments

The gratitude of the author is due to William Voglesong of PSI Associates, George W. Leyda of 3M Company, James R. Cox of Cosar Corporation, and Norman W. Scharpf of the Graphic Communications Association for their editorial contributions and suggestions to this publication.

The industry's thanks are due to all other members of GCA's Densitometry Task Force for their encouragement and enthusiastic support throughout the difficult and often frustrating T-Ref development period. Led initially by Howard Cutler, then of Foote, Cone & Belding, and later by Robert Miller of Liberty Engraving, other members of the Task Force include James Cox of Cosar, Ken Smith of DuPont, Charles Rinehart of Eastman Kodak, Paul Borth of the International Prepress Association, Robert Van Arsdell of MacBeth Process Measurement, Glenn Frazier of Meredith Corporation, Roy Berns of the Munsell Laboratory, George Leyda of 3M, Tom Lumenello of Polaroid Corporation, Milt Pearson and Paul Swift of the RIT Research Corporation, Miles Southworth of the RIT School of Printing and Chair of the Gravure Association of America (GAA) Densitometer Committee, David Crowley of Tobias Associates, David Bowden of the X-Rite Company, and Norman Scharpf and Peter Brehm of the Graphic Communications Association.

Special thanks are due Milton Pearson and Paul F. Swift of the Rochester Institute of Technology (RIT) Research Corporation, Roy S. Berns of the RIT Munsell Color Science Laboratory, Thomas Lumenello of Polaroid Corporation, and William Voglesong of PSI Associates, all of whom have contributed their time and resources to bring the T-Ref program to fruition.

This book is dedicated to Franc A. Grum for his vision and support of the T-Ref project.

About GCA

The need for coordination and communication between all links in the graphics chain is essential for continued productivity enhancement in the industry. The Graphic Communications Association, with its broad ranging membership of publishers, color separators, printers, advertising agencies, manufacturers, and suppliers, stands in a unique position to facilitate this coordination. Through its conferences, publications, and committee research work, GCA has, since 1967, brought together representatives of the graphic communications industry to help meet their productivity goals. Membership is open to all industry firms and organizations. Your inquiry is invited.

Table of Contents

Foreword ix

I.   **The Basics of Densitometry** 1
What is a densitometer? 1
What are the uses of reflection densitometers? 1
How are reflection densitometers used in prepress? 2
How are reflection densitometers used in the pressroom? 2

II.   **How Densitometers Work** 5
Do densitometers measure density? 5
How is reflection density calculated? 6
What are the components of a densitometer? 7
What role do standards have in how densitometers work? 8
What is the relationship between illumination angles and glossy
    surfaces? 8
What is a densitometer's "response"? 9

III.   **Wide-band, Narrow-band, and Band-width** 11
How do wide-band and narrow-band densitometers differ? 11
Why are wide-band densitometers used? 13
Why are narrow-band densitometers used? 14
Which densitometer should I use? 16

IV.   **Polarized Filters Explained** 17
What are polarized filters, and what are they for? 17

V.   **Using Densitometers** 19
Should I "zero" on the paper? 19
Should I measure paper density and subtract it out when
    measuring/calculating print attributes? 19
What is the proper backing when measuring printed sheets, proofs,
    off-press proofs and other materials? 20
What else do I need to know about backing? 21
Can I measure dot area on plates with a densitometer? 23
How do I use a densitometer to measure dot gain? 24
Are densitometers "colorblind"? 25
Should I use a densitometer or a colorimeter? 26
What is the role of color bars in print control? 26

If I can't fit a color bar on my form, what should I do? 27
What is print contrast, and why is it important? 28
Which densitometer is best for my operation? .29
What should I ask when buying a densitometer? 29
How sensitive are densitometers to surface irregularities? 31

VI.   **Improvements in Densitometry** .33
        What improvements have occurred in densitometry? 33

VII.   **Standards and Verifying Conformance to Standards** 37
        About the ANSI/ISO standard 37
        About the DIN standard 38

APPENDICES

Appendix A: The Process of Using the GCA T-Ref™ 47
        What is the GCA T-Ref? 47
        How are T-Refs calibrated? 48
        How do T-Refs fit into a production setting? 48
        How close should my densitometer's readings be to the T-Ref values? . .48
        How close are T-Ref's readings to the ANSI/ISO standard? 49
        How does T-Ref differ from the SWOP Hi-Lo Reference? 49
        What does SWOP say about Status T and T-Ref? 49
        How often do I use T-Ref? 49
        Can T-Ref be used with any kind of work? .50
        Do I calibrate on T-Ref? 50
        What happens if I calibrate equipment on T-Ref color patches? 51
        Why do I need to replace T-Ref? 51
        What's a good way to use T-Ref in my operation? 52
        Can variation among Status T densitometers be greater than ± 0.02? 53
        Is improvement possible? 54
        What is "total uncertainty"? 54
        Summary: it's good 56

Appendix B: Standard Lighting 57
        Why we use standard lighting 57
        "Standard" lighting may not be 57

**Appendix C: Standard Methodologies for Determining Densitometry
  Capability** 59
        To understand variation caused by the densitometer and operator
           together 59
        To understand variation caused by the densitometer alone 61

To understand variation caused by the environment 62
To understand variation caused by the light in the surrounding
  environment 64
To understand variation caused by the densitometer's photometry 65
To understand when a unit should be recalibrated 65
To understand variation caused by aperture size 66

Appendix D: Common Characteristics of Densitometers,
  Colorimeters, and Spectrophotometers 71
All three share a common measurement approach
  and geometry 71
Spectral response: different slices of the same light 72
Signal processing: one approach, different data formats 75
Applications 76

Foreword

The need for improved communication of density values has been recognized for many years. According to noted educator, author, and consultant Miles Southworth in his article, "Densitometers Are Better Than You Think!," published in the *GTA Bulletin*, (Summer, 1973, Vol. XXIV, No. 2, Page 1):

> The densitometer is one of the most important and widely used instruments in the graphic arts today. It is no longer a luxury. It is a necessity. Any attempt to control production without it would be hopeless. The densitometer has become the lifeblood of the quality control department everywhere. Since the quality level, as well as the profit and loss, of a printer hinges on decisions that are based on numbers gained from using a densitometer, it should be obvious that [the densitometer] should be as accurate as possible. More important, however, is that each instrument should [provide readings that are able to be compared with readings taken by] every other instrument. Printers need to be able to confidently compare readings, exchange information, and know they are speaking the same language.

Of course, it is not only the printer for whom the densitometer is necessary for quality control. It is the advertising agency, the publisher, the cataloger, and the color separator as well. Unfortunately, inter-instrument agreement was not the rule in 1973. The lack of a standard response, coupled with differing manufacturer interpretations of the correct response, created communication problems between and among each segment of the print production community and caused doubt about whether densitometric values could be relied upon to provide needed process information.

The basis for a solution to the lack of densitometer agreement was finally achieved in 1984 when the American National Standards Institute (ANSI) Committee PH2 (now IT2) finalized a spectral response for graphic arts reflection equipment, naming it Status T. By specifying a response instead of a filter set in this ANSI standard, the group acknowledged that several different combinations of filters, optics, and internal electronics could achieve the same result, and that the marketplace would dictate what these combinations might be. The need was, and is, for all wide-band reflection equipment to read a given density and report a common number, regardless of the means used to get to that common number. This approach was also adopted as an international standard with the publication of ISO 5/3 by the International Organization for Standardization (ISO).

Although the Status T response gave the graphic arts industry a common standard, what was still lacking was a way to apply this standard in a practical production setting. No means for verifying the conformance of the densitometer to the standard existed—until T-Ref was developed by the GCA Densitometry Task Force. Proposed to the Prepress Committee by James Cox of Cosar, development of the T-Ref—including the design, production, and spectrophotometric calibrations—was initially undertaken by Franc Grum of the Munsell Color Science Laboratory under contract to the Graphic Communications Association through the RIT Research Corporation before his untimely death in 1986. His assistant, Roy Berns, carried on this development effort with the full cooperation of GCA's Task Force, which numbered among its members the industry's leading densitometer manufacturers. William Voglesong, formerly head of one of Eastman Kodak's Central Densitometry Labs and now president of PSI Associates, ultimately provided the necessary insights into needed calibration refinements.

Since its unveiling, T-Ref has proved to be an effective means of verifying wide-band reflection densitometer conformance to the ANSI/ISO standard. By providing an independent verification of the equipment's response, T-Ref enables users to communicate by numbers, knowing that any differences that arise are due to the proofs and press sheets that they are measuring and not to differences in densitometer values.

With T-Ref's development completed, the Task Force has perceived the need for wider understanding of basic densitometric terms and approaches, including the developments that have occurred in this technology over the last five years. Together with GCA Publication 205, a technical reference to T-Ref's calibration methodology, this present document is intended to meet that need. To that end, the GCA Densitometry Task Force welcomes comments and suggestions about these contents.

I.

The Basics of Densitometry

What is a densitometer?

Densitometers are instruments designed to determine, indirectly, the light absorbed by a surface. They do this by comparing the intensity of the light that reflects off (or transmits through) a surface to the intensity of the light shown upon it, and then calculate density via an accepted logical relationship.

There are two kinds of densitometers:

- **Transmission densitometers** measure the amount of light that is transmitted through a transparent material such as a film base.

- **Reflection densitometers**—the subject of this publication—measure the amount of light reflected from a print and are a critical aid in quality control for all involved in the printing production process.

What are the uses of reflection densitometers?

In printing and publishing, reflection densitometers are used extensively in prepress and pressroom operations by advertising agencies, publishers, color separators, printers, and suppliers for greater quality control. By measuring the ratio of reflected to incident (source) light of the red-green-blue components of light, densitometers provide a direct reading of optical reflectances (the reflectance of light off the substrate and colorant, such as the ink or dye). Based upon these reflectance values, reflection densitometers can also calculate key print attributes, such as major filter density, contrast, and dot gain, as well as be used to infer properties such as hue error.

How are reflection densitometers used in prepress?

In prepress, reflection densitometers can be used to:

- Measure the highlight, middle-tone, and shadow density regions of photographic prints and artwork that are ready for halftone reproduction. Measuring the lightest and darkest areas of these materials provides information that will help predict the exposure necessary to properly create an image on the printing plates that matches the range achievable by any given ink-paper-press condition.

- Specify the colors and the range of illumination for a subject being photographed. This data can be used by the layout artist and even the studio photographer to make adjustments to improve the final printed product.

- Analyze press proof and off-press proof characteristics to control the variation of color and tone reproduction.

- Analyze the incoming materials, such as ink and paper, used in proofing and off-press proofing.

- Discover variations in proof press printing.

- Determine masking factors.

- Determine photographic emulsion characteristics (also called sensitometry).

How are reflection densitometers used in the pressroom?

In pressroom areas, reflection densitometers can be used to:

- Analyze the quality of supplied off-press proofs and press proofs and determine their conformance to standards and specifications.

- Analyze the incoming materials, such as ink and paper, used in the pressroom.

- Determine how a press is performing.* Densitometers can be used to evaluate print characteristics such as color consistency from sheet-to-sheet; color uniformity across the sheet; the amount of dot gain and slur; relative ink film thickness; and color matching of the proof or OK sheet relative to the production printing. This information can be used for both trouble-shooting and statistical process control.

- Analyze color bars to gain information to adjust ink laydown, monitor ink and fountain solution consistency, and check the condition of the plates in terms of blinding and scumming.

Throughout the printing process, densitometers allow the print professional to communicate process information based on objective numbers rather than subjective adjectives. Numeric values are key to effective communication about the quality of the printed product and the control of the production process. For such applications, the density values must be based on a common definition, such as Status T.

---

* **Caution:** Note that measurements made during a press run should be viewed as a snapshot of the operation and that control actions should be made only when a definite trend is noted. Short-term, cyclical changes may be traced to instability in the blanket or a batch of ink or fountain solution, or a number of other temporary conditions. Long-term changes or drifts in the printed results may be traced to such conditions as variations in paper absorption, plate wear, or press adjustments. Contact the Graphic Communications Association for additional literature about application of statistical process control (SPC) tools and the broader Quality Process Commitment (QPC) Program needed for effective implementation of these tools.

II.

How Densitometers Work

Do densitometers measure density?

Densitometers do not, as their name suggests, actually measure density. They instead measure reflected light, assume that the light absorbed is the difference between the reflected light and the light the densitometer supplies, and then calculate the density via an accepted logarithmic relationship.

Density can be defined as a measure of the proportion of incoming light absorbed by a surface. However, the amount of light absorbed by the surface is a very

DIAGRAM 1: *Light hitting surface and reflecting to eye.*

difficult thing to measure with any but the most elaborate and expensive equipment. Therefore, the accepted procedure is to measure the proportion of light reflected from a measured surface and to assume that—for practical purposes—the amount of light absorbed is equal to the amount of incident light (the supplied light) minus the amount of light reflected.

How is reflection density calculated?

Density is calculated by densitometers using the equation:

$$Density = log_{10} \, 1/R \text{ (where R = reflectance)}$$

This is a pure definition, which means that it defines the relationship between light reflectance and density. This definition numerically interprets density, and,

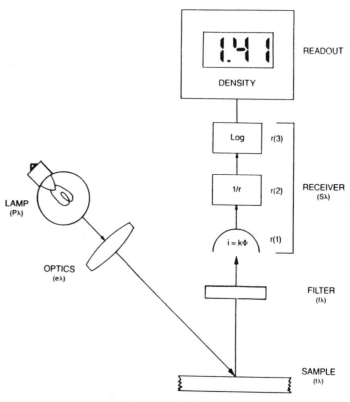

DIAGRAM 2: *Schematic of a densitometer.*

more importantly, describes density in a manner that approximates the way in which the human eye sees objects (see Diagram 1).

What are the components of a densitometer?

To accomplish this measurement and calculation, a densitometer employs three essential components— and, of course, a sample to be measured—and a means for analyzing the measured light, i.e., a computer. The three components (see Diagram 2) include:

1. **An illumination system** consisting of a lamp, illumination optics, and a power supply for lamp operation. By convention, and in keeping with ANSI/ISO standards, this lamp provides light with a color quality (i.e., a balance of red, green, and blue) called Illuminant A. This color quality is very close to the quality of light emitted from an unfiltered tungsten lamp. The densitometer lamp is powered by an electrical circuit that either keeps it at a well controlled brightness or pulses the lamp so that the quantity of light per flash is constant. Regardless of the light's power source, however, the lamp and power supply assure a known illumination for the incident portion of the ratio.

2. **A collection and measurement system** consisting of a photoreceiver, light collection optics, and spectral trimming filters that transmit to the photoreceiver only certain portions of the full visible spectrum while blocking other portions. This collection system usually contains color filters to provide for an overall spectral sensitivity matching some stated standard.*

    Densitometers used in printing and publishing have historically used a filter set that approximated the color separation characteristics of the printing process. These spectral values were refined by the ANSI IT2 Committee to produce a documentary standard called Status T. This engineering specifica-

---

* Which standard the densitometer responds to is a vital element in evaluating its application. For example, when the filter-photocell spectral response matches the luminosity (brightness irrespective of color) response of the human eye, the densities are said to be "visual." Other spectral responses make the densitometer an abridged spectrophotometer or an analogous colorimeter, defined as a color measurement device that has all the color measurement capabilities of a colorimeter. As such, the densitometer can measure color like a colorimeter, although when used in this manner it provides output in terms different than the L*a*b* terms used by a colorimeter.

tion has made it possible to match densitometers produced by the same manufacturer, as well as models produced by various manufacturers. Status T density specification is the result of design, measurement, and certification aimed at meeting a standard response definition.

3. **A signal processing system** that takes the electrical signals representing incident and collected light energy and provides the computation and display function. This system may be a simple ratio detector and logging circuit connected to an analog or digital display, or it may include memory storage, providing for the processing of derived functions such as dot gain and contrast.

What role do standards have in how densitometers work?

The complex interrelatedness of a densitometer's components underscores the need for standards that define its capabilities and 'response' characteristics. The American National Standards Institute (ANSI) and the International Organization for Standardization (ISO) have agreed upon standards that place some design constraints on this instrumentation. For example, the illuminating and collecting optical systems are limited to a 5 degree cone angle, and the optical axes of inclination to the sample are measured from a line perpendicular to the surface. By standard agreement and practice for at least the last forty years, all reflection densitometers have been manufactured with an illumination angle of 45 degrees or 0 degrees. Thus, the sampling of the reflected light is taken at 0 degrees or 45 degrees.

What's the relationship between illumination angles and glossy surfaces?

In the case of smooth or glossy surfaces, a portion of the incident light is reflected at the same angle that it hits the surface. This light is called *specular reflection* and is not picked up by the light detectors located at the 45-degree angle in densitometers. When measuring smooth and glossy surfaces with a densitometer having light collection sensors at 45 degrees, readings indicate that such surfaces are absorbing more light—give higher readings—than is actually the case. This specular reflection is detected by the human observer as gloss, but only at a particular angle of viewing. While the observer is aware of this phenomenon, **he or she will invariably position the printed surface at an angle that eliminates this specular reflection, or gloss. Thus, for normal viewing and evaluation of the printed surface, the densitometric measurement agrees quite well with the human perception that the image appears denser.** In fact, varnish is often used to achieve this dense appearance of an image. The varnish accomplishes this by giving the sheet a smooth surface that creates a lot of specular reflection.

What is a densitometer's "response"?

All densitometers have built into them the means to determine a density value using the log relationship. Densitometers may give different readings if any component in their respective measurement systems—such as the filters, light sensor, or the log relationship built into the unit—is different. The densitometer's *response* is defined as the density readings provided by that unit based upon all of the input variables required to achieve those readings. Thus, a *response* standard is one that requires all densitometers to provide uniform density readings regardless of the filter sets, light sensing devices, or log relationship built into the unit.

III.

Wide-band, Narrow-band, and Band-width

The term "band"—a shorthand reference to "band-width" or "pass-band"—refers to the range of the visible spectrum that a filter will allow to pass through the filter. Wide-band filters allow a wider range of light to pass through than do narrow-band filters. A more precise understanding is required, however, to fully appreciate differences between these filters.

How do wide-band and narrow-band densitometers differ?

To understand in general how wide-band and narrow-band densitometers differ, it is important to keep in mind that **a densitometer is used primarily to give an objective, numerical measurement of a surface that is related to the subjective view that a human makes when looking at the same surface.**

Recall that a reflection densitometer measures the **reflectance** of a surface, usually differentiating the kind, or color, of the incident light into three portions of the visible spectrum. These three portions generally conform to the visible perception of red, green, and blue light.

Human vision allows us to see light in a visible spectrum that ranges between 400 and 700 nanometers (nm), a nanometer being a measurement of wavelength (see Diagram 3). The spectral region we see as "blue" spans approximately the range 400nm to 500nm, while 450nm to 640nm is "green," and 500nm to 700nm is

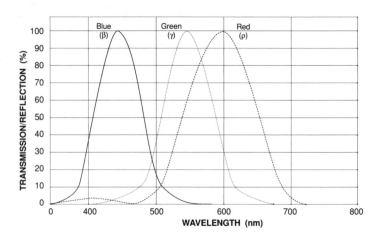

DIAGRAM 3: *The probable sensitivities of the three color mechanisms of the eye. All values have been adjusted so the peaks are 100%.*

"red." Like the human eye's red-green-blue receptors, wide-band filters each transmit roughly 100 nanometers, or one-third, of the visible 300-nanometer range. Wide-band filters are also similar to filters used in color scanners for color separation production. Narrow-band filters, which are usually interference filters, have approximately a 20nm band-width. The spectral specification of narrow-band densitometry requires a measurement of the peak response, the band-width at 50% response, and, most importantly, some measure of the rejection (lack of response) outside the pass-band.

In Europe, the industry has for many years used Wratten 47B filters for measuring the density of yellow inks. The 47B filter measures a band-width of approximately 55-60nm. In the United States, the industry has chosen instead to use the Wratten 47 filter, with a band-width of approximately 70-80nm, to measure yellow ink. The United States has standardized, through the American National Standards Institute (ANSI), on a response standard that is premised on the industry's choice of the Wratten 47 filter. The 47B filter gives higher readings in yellow than the 47 filter, and sometimes densitometers using 47B filters have been referred to as "narrow-band densitometers." They are, however, more appropriately classified as wide-band densitometers (recall that narrow-band filters typically

Type of Filter	Band-Width*

To measure yellow ink:

The Blue mechanism of the eye sees 400-500 nanometers	
Wide-band 47	*measures 400-510 nanometers*
Wide-band 47B	*measures 400-480 nanometers*
Narrow-band	*measures 415-445 nanometers*

To measure magenta ink:

The Green mechanism of the eye sees 450-640 nanometers	
Wide-band 58	*measures 490-570 nanometers*
Wide-band 61	*measures 500-560 nanometers*
Narrow-band	*measures 515-545 nanometers*

To measure cyan ink:

The Red mechanism of the eye sees 500-700 nanometers	
Wide-band 25	*measures 580-640 nanometers*
Wide-band 29	*measures 590-640 nanometers*
Narrow-band	*measures 605-630 nanometers*

*All values are approximate; see Diagrams 4A, 4B, and 4C for graphical details.

have a substantially narrower, 20nm band-width) and will be considered wide-band in this document.

Why are wide-band densitometers used?

Wide-band densitometers provide functional measurements. A functional measurement is one that indicates how a sample will "look." Wide-band measurements usually correspond to some attribute of the printing system, such as ink film thickness, laydown, print contrast, exposure, dot gain, and so on. Wide-band units provide measurements that may be considered comparable to colorimetry, exposure metering, and other measuring technologies when they are used to evaluate process color reproduction.

Further, human eye sensitivity to the three portions of the visible spectrum is similar, although not identical, to that of a wide-band densitometer (see Figure A). This correlation is, in fact, quite good in the context of any discussion focused on four-color reproduction in printing and publishing.

Why are narrow-band densitometers used?

In contrast to wide-band densitometers, narrow-band densitometers have a spectral pass-band of approximately 20nm. If these spectral zones are centered on the wavelength of maximum absorption, narrow-band units provide the greatest sensitivity of measurement. As such, these measurements are said to be analytical because they indicate how much absorber is present at these selected wavelengths of maximum absorption. Narrow-band densitometers, however, have dead zones between their pass-bands. In their dead zones, these units are not sensitive to anything in the sample being measured. One result of this is that two inks that the narrow-band units perceive as identical within the 50nm band could, if their spectral characteristics are different in the dead zones, appear different to the human eye. In order to relate narrow-band measurements to wide-band measurements, the user must know about the spectral characteristics of the sample being measured.

Narrow-band filters are preferred by some because their measurements produce larger numeric differences for a given change in ink-film thickness. Measurement systems using narrow-band filters are also less dependent on lamp and photoreceiver variation. However, tests have shown that the signal-to-noise ratio in narrow-band units—the amount of light that a narrow-

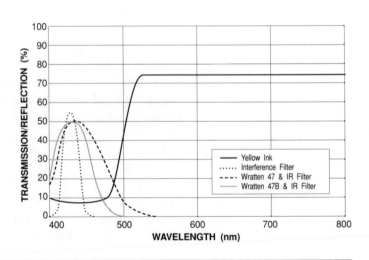

DIAGRAM 4A: *Spectral reflectance of a yellow solid print and transmission of color filters (infra-red trimming filters included).*

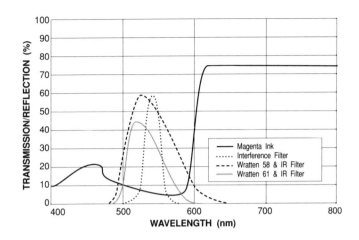

DIAGRAM 4B: *Spectral reflectance of a magenta solid print and transmission of color filters (infra-red trimming filters included)*

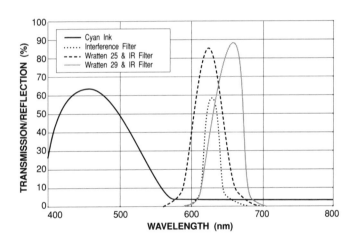

DIAGRAM 4C: *Spectral reflectance of a cyan solid print and transmission of color filters (infra-red trimming filters included).*

band densitometer's signal processing system must sense in order to display a genuine change in density— is not as good as the signal-to-noise ratio in wide-band equipment. Also, the increased sensitivity in the areas of band-pass sensitivity is gained at the expense of being insensitive to sample variation taking place in the dead zones between narrow-band regions of sensitivity. As such, densitometers using these narrow-band filters do not relate well to the human observer.

Which densitometer should I use?

Through improved standards and electronics, wide-band densitometers now have the sensitivity to record small changes in light in a repeatable manner. Though experts have suggested that there are areas of manufacturing—especially where the measured color of the product being manufactured is known—where narrow-band densitometry could have application, they suggest that printing, publishing, packaging, and other measurement and control applications could suffer using narrow-band equipment because of the data that is not being captured. These experts recommend using wide-band equipment.

IV. Polarized Filters Explained

What are polarized filters, and what are they for?

In the late 1950s and early 1960s, the Graphic Arts Technical Foundation (GATF) investigated the addition of light-polarizing filters to reflection densitometers for the measurement of wet inks. The intent of this innovation was to allow measurement of wet ink in a way that wet ink and dry ink would read the same (recall that glossy or smooth surfaces, like that of wet ink, produce specular reflections and thus higher density values).

Polarized filters sometimes produce reductions in density similar to that which occurs during dry back. They are, however, ineffective in predicting dry ink film density because:

- Wet ink dries to have varying degrees of gloss, so that any repeatable correlation between the wet and dry ink films is not possible.

- Dry samples that have different gloss and differing appearances to the human eye may read the same.

Polarized light is known to have the effect of eliminating portions of specular reflections. It was hoped when developed that polarized filters would mimic the specular reflection, or glossy component, of printed inks and make the densitometer measure dry inks as if they were wet. However, since the $0°/45°$ densitometer itself eliminates, or at least minimizes, detection of any specular reflection, the use of polarizing filters in such a densitometer is unnecessary as well as ineffective.

Printers are frequently led to believe that the polarizing densitometer will predict the result of dry-back—a phenomenon where wet ink is absorbed and loses some or all of its gloss, thus lowering its effective density. In fact, the polarizing densitometer fails to predict the results of dry-back.

V.

Using Densitometers

Should I "zero" on the paper?

No. To "zero" on the paper means to measure the density of a sample of the unprinted paper being used and adjust the densitometer to automatically subtract this paper density out of subsequent readings. Since paper density and paper color are important components in the reproduction of an image, they should be included in the density measurement. Therefore, the densitometer should *not* be zeroed on the paper.

Should I measure the paper density and subtract it out when measuring/ calculating print attributes?

The answer depends upon the information you want:

- **Yes**, if you are interested in finding out the effect that the ink or other colorant *alone* is having on the visual appearance of the reproduction.

- **No**, if you want to learn more about the *total* visual impact of the paper plus ink or other colorant because paper has a density and color that contribute to the visual appearance of an image. The print contrast calculation is an example of where you would want the total visual impact of paper and ink.

- **Yes**, if you are making a print attribute calculation that by convention subtracts out the paper density. Dot gain calculation and trap calculation are examples where paper density is subtracted. (Note: By convention, most, if not all, modern densitometers automatically subtract out paper density when making dot gain calculations.)

It is recommended that all densitometer users— print buyers, separators, and printers—follow this guideline when making densitometer readings to assure a good understanding of densitometric values.

What is the proper backing when measuring printed sheets, proofs, off-press proofs, and other materials?

The proper backing when measuring printed sheets, proofs, off-press proofs, and other materials is black, not gray or white. *ANSI/ISO 5/4-1983—ANSI PH2.17-1985: Density Measurements Geometric Conditions for Reflection Densitometry* defines geometric conditions for reflection density measurements. Included in this standard [note: the reflection density value is being revised to include the ± 0.20 range] is the statement:

> It is important to recognize that this International standard specifies that the surface behind the specimen shall be spectrally non-selective, diffuse reflecting, and has an ISO reflection density of 1.50 ± 0.20. [Note: a spectrally non-selective surface is gray; a gray tone with a reflection density approximating 1.50 would appear black.] Some reflection density standards have generally specified backing materials with much lower reflection density." Annex A of the standard explains further: "It is necessary to specify the characteristics of the material used behind a specimen when determining reflection density to define unequivocally the measurement.

According to the revised document, the choice of a black backing was made for several reasons, including:

1. The black backing reduces measurement variability. This is important for metrology purposes as well as process control purposes since many specimens used in printing and publishing are generally not totally opaque, and printing on the reverse of a sheet will affect measurements, sometimes substantially. Research by PSI Associates indicates, in fact, that measurements of substrates with a transmission density of 1.00 or lower would be affected by dif-

ferent backings and printing on the reverse side of a sheet.

2. A low-density surface greatly reduces the problems associated with maintaining the backing surface from the standpoint of spectral, density, and physical requirements.

3. Using a black backing permits the calculation of absorptance factor directly from the density readings.

The standard also cautions that in the printing industry specimens are often viewed by backing them with two or more layers of the same substrate. It suggests that there is no reason why ISO densities of a specimen consisting of a combination of an image-bearing material along with substrate cannot be read together, although precaution should be taken to identify the readings as being from the combination of the image layer and substrate and not the image layer by itself.

What else do I need to know about backing?

The impact on measurement variation resulting from not using standard backing materials increases as papers get thinner and transmission densities decrease. Initial research by PSI Associates into this topic indicates that for graphic arts papers, the delta density of the paper as measured with white and black backing varies from 0.07 down to 0.002 as the transmission density of the support varies from 0.3 up to 1.2, with data indicating an exponential relationship between substrate transmission density and the difference to the measurement values of that same substrate taken with a black and white backing. The white used for this research was a non-selective white ceramic-on-steel tile with no fluorescence and a reflectance of 89%; the black backing was the GCA BackStop™ conforming to ANSI/ISO requirements. To provide practical insight into transmission density values of paper, PSI evaluated several sheets, discovering that thin papers, such as those used for multi-part forms or bibles have a transmission density of approximately 0.30. The transmission density for 20 pound white wove offset is approximately 0.60; for 80 pound coated card stock it is approximately 0.90 and for 120 pound card stock it's approximately 1.20. This research indicates as well that photographic papers follow a similar pattern, although have a different curve.

Understanding that a black backing minimizes measurement variation, use of the standard backing is necessary in the following situations:

- Instrument set-up

- Instrument calibration (note that it is essential that calibrating targets used for a densitometer set-up be backed with black because such backing is part of the definition of ANSI/ISO density)

- Job specification (e.g., printing a job to specified aim points), unless a different backing is specified

- Industry specification (e.g., printing types of products, such as magazine advertisements, to specified aim points), unless a different backing is specified

- Communication of measurement values between organizations or physical locations, unless a different backing is specified

For internal uses, such as when an operator is subjectively evaluating an image as compared to a proof or another press sheet, a black backing will give the printed image a slightly different appearance and is probably less applicable. As the standard already suggests, the operator may in these cases want to place one or more pieces of unprinted substrate under the sheet.

Several members of the international community have proposed development of a *standard* white backing. Specification of such a backing would necessarily entail defining the following parameters:

- Color

- Brightness

- Surface diffusion

- Fluorescence

- Density

For purposes of measurement communication, use of a white backing would require recording of each of these factors so that recipients of the measurements would understand the context in which such measurement were made.

The robustness of the current standard from both metrology and user standpoints, the simplicity of using a black backing (materials such as Backstop™ exist and are in use that allow conformance to the current ANSI/ISO standard), and the challenges associated with defining a white backing suggest that development of a standard white backing would not help minimize measurement variation and user community application of a standard backing material.

In summary, the industry should use a black backing when making color measurements for metrology and process control purposes. This should be the standard approach so that these measurements are not influenced by printing on the reverse side of the sheet or by factors such as images on sheets beneath the sheet being measured; measurements should not be made from sheets stacked on one another. Above all, if densitometer measurements are made without a black backing, users should communicate the backing that was used and also be aware that these measurements will be subject to additional variation. How much variation will be contributed can be determined by comparing readings of the same sheet taken with and without the standard black background.

Can I measure dot area on plates with a densitometer?

You can, but the repeatability and accuracy of the data are questionable. Recall that densitometers calculate density information based on reflectance of light. Many plates in use today do not use emulsion colorants that are dark enough, nor plate colors that are light enough, to provide reflected light with enough intensity to yield repeatable information. You may wish to conduct tests comparing densitometric values with actual measured dot values in order to determine this correlation.

How do I use a densitometer to measure dot gain?

Dot gain, often called apparent dot gain, encompasses both mechanical (the growth of the physical halftone dot) and optical (how the physical halftone dot appears to the human eye) elements.

Apparent dot gain can only be calculated by knowing both the supplied or target film area and the apparent dot area (also called the equivalent dot area) of the same target or area on the printed sheet:

Apparent Dot Area in Print - Film Dot Area = Apparent Dot Gain

Apparent dot area in the print may be measured using either the automatic function available on many densitometers or manually using the following values of reflection density:

- Density of the paper background of the print

- Density of the solid patch of color located near the target area, such as the 50% middle-tone tint

- Density of the printed tint

Apparent dot area is then calculated using the Murray-Davies equation:

$$\%ADA = \frac{1 - 10^{-D_t}}{1 - 10^{-D_s}} \times 10$$

D_t = *Density of the printed tint minus the density of the paper.*
D_s = *Density of the solid patch minus the density of the paper.*
ADA = Apparent dot area in percent.

By convention, total dot gain is an incremental, or add-on, increase in apparent dot area. As an example, a printed 50% tint that, using the Murray-Davies equation, measures as having a 72% total dot area, has a total dot gain of 22%. Thus, 22% dot gain **does not mean** that a 50% dot has increased by 22% and is now measuring like a 61% dot (see Diagram 5).

DIAGRAM 5: *Dot gain measurement is incremental.*

$$50\% + 22\% = 61\%$$

$$50\% + 22\% = 72\%$$

Are densitometers "colorblind"?

No—in fact, wideband Status T densitometers are often used to identify colors precisely. Here's how:

Every ink printed—pastels, bright reds, dark greens—will absorb and reflect light based on the cyan, magenta, yellow, and black inks used to create that overprint. By measuring the overprint solid or tint, using all four of their wide-band densitometer filters (visual plus red, green, and blue), users can calculate the overprint's major filter density (the filter that yields the highest density reading), minor filter density (the filter that yields the lowest density reading), and the major-minor filter density (the filter that yields the middle density reading). By recording these values and using simple equations to calculate hue error and grayness that can be plotted on the GATF Color Circle, users will know the color's major filter density, hue, and grayness and can identify that color uniquely using numbers derived from a densitometer.

How does this differ from a colorimeter? Instead of using density values, a colorimeter identifies a color using L*a*b* values, or some similar measurement system used to define color outputs.

Should I use a densitometer or a colorimeter?

Some of the factors that users need to consider in deciding whether to use a densitometer or a colorimeter are:

- **Cost:** Densitometers are less expensive.

- **Standards:** Status T defines the standard wide-band densitometric response (and the T-Ref serves as a measurement system to check conformance to this standard—see Appendix A). Colorimetry is well defined by CIE response functions, but several reporting systems (e.g., L*a*b*, L*u*v*, and L*c*h*) are used, with no one system accepted as a universal standard and no user-friendly method existing to check conformance to any one approach. Thus, one user's colorimeter could identify a measured color using L*a*b* values that could be different from L*a*b* values determined by a second user measuring the same color patch with a different colorimeter.

- **Application:** For a printer measuring sheet-to-sheet consistency, or proof-to-press correlation, a densitometer is probably the best solution. For the separator correlating a printed proof with a piece of fabric or a transparency, or for a printer confronted with the same task, a colorimeter is best since the gamut, or range, of colors achievable using clothing or food dyes is distinct from that possible using printing ink pigments, suggesting the need for a common color definition method.

For a complete outline of how densitometers, colorimeters, and spectrophotometers compare and contrast, see Appendix D.

What is the role of color bars in print control?

The human eye can perceive 1,000,000 colors. Process inks—cyan, magenta, yellow—plus black can create approximately 4000 colors. Although a densitometer can measure all 4000 of these colors, such measurement is difficult and expensive. Color bars allow the print professional to measure selected key print characteristics in order to simplify color reproduction measurement and control.

Here's how color bar patches simplify the process:

- Measuring the cyan-magenta-yellow-black ink densities reveals information about these four foundation inks (not considering special spot colors) that are used to reproduce all colors.

- Measuring the solid overprint red-green-blue tells you how well the inks are trapping throughout the image by examining how well they are trapping on these patches. Keep in mind that solid overprints are worst-case examples; most halftone areas offer some paper onto which second-, third-, and fourth-down inks can adhere.

- Measuring the 25%, 50%, and 75% tint areas for the cyan, magenta, yellow, and black printers allows you to know the dot gain in these tint areas. When considered with the solid ink densities and paper densities, this information provides us with knowledge about the tone reproduction of our printing press system, which includes press, ink, paper, operators, film, plates, blankets, and other factors.

By measuring—and analyzing—these few color bar patches, we can better understand and control the entire printed reproduction in the most cost-effective manner.

If I can't fit a color bar on my form, what should I do?

Experts suggest that, at a minimum, printers should try to include a solid patch plus adjacent 75% tint for each printed color—as well as for black—somewhere on the form—even if it has to be put in the gutter. This is recommended since we perceive good printing—printing that has "snap"—to be open in the shadow region of the printed reproduction. A measurement of print contrast, which is taken using the solid and 75% tint, allows us to use a number to define this print contrast.

If using a color bar on every printed job is not possible, an option is to use a color bar on as many jobs as possible, or on a regular basis, such as once a month. Consider also running a test form on a regular basis to track press performance. It will allow you to diagnose reproduction problems before your customer has the chance to do it for you while evaluating a live job. Remember: some measurement is better than no measurement.

What is print contrast, and why is it important?

Print contrast is a measure of shadow contrast, which is the degree to which viewers can distinguish printed tones in the shadow areas of a reproduction. Print contrast is calculated in a manner that compares density reading differences between a three-quartertone tint area (usually a 75% tint) and a solid patch. The formula is:

$$\%PC = \frac{D_s - D_t}{D_s} \times 100$$

D_s = Density of the solid patch (including the density of the paper).
D_t = Density of the printed three-quartertone patch (including the density of the paper).
$\%PC$ = Print contrast in terms of percentage, e.g., 25%.

Research by GCA's Print Properties Committee has suggested that when printing on a No. 5 coated groundwood stock using the heatset web offset process, that a print contrast value of 25% or higher corresponds well with what viewers perceive as good printing.

Print contrast is thus a useful print attribute to measure because:

- Printers and separators find it useful since it provides an indication of the image's tone reproduction at an important point in the tone curve.

- Print buyers—whether or not they realize that they are visually evaluating print contrast—use print contrast as an indication of print quality. Printed sheets with high print contrast are sometimes referred to as having "snap" since the rich, deep shadow areas and open middle-tone and three-quartertone areas appear to have a three-dimensional characteristic and thus "snap off" the page.

A print contrast number by itself provides a wealth of process information, such as 1) the density of the ink laydown, 2) the dot gain in the three-quartertone area, and 3) the capability of the printing process to increase ink density without gaining in the middle-tones of the reproduction. For example, if the print contrast value increases, it suggests that the solid ink density has increased without an increase in the density—which would mean dot gain—of the three-quartertone patch, increasing the visual contrast between the solid and the three-quartertone areas. Likewise, a drop in print contrast suggests that as density increased, the three-quartertone area filled in, flattening the image reproduction and decreasing the visual contrast between the solid and the three-quartertone areas.

Which densitometer is best for my operation?

A *modern* densitometer is best for your operation. Units manufactured within approximately the last five years (i.e., since 1985) usually qualify as modern. A modern unit has many, if not all, of the following features:

- Solid-state electronics (if a unit uses vacuum tubes, replace it; these tubes fluctuate).

- Digital read-out (if a unit uses an analog meter to record density values, replace it since operator interpretation of analog readings is subject to variation).

- Conformance to ANSI/ISO geometry and spectral specifications.

- Easy calibration or standardization procedures.

- Computer communication links available via an EIA RS-232 port.

- Use of regulated or pulsed halogen lamp.

- Battery operation or long, flexible, power cord.

- If it is a portable unit, it should be portable enough to move and use around the plant without strain.

What should I ask when buying a densitometer?

When discussing densitometry equipment purchases with a manufacturer's representative, you will find answers to the following questions helpful—in addition to the information listed above—for comparing equipment:

- How many readings can the unit make in a normal battery-charge cycle?

- Do densitometer readings change near the end of the charge cycle?

- Is the light source the standard Illuminant A 2854K? (Although you will not verify equipment conformance to the standard, you can see any gross difference by comparing the densitometer light source visually with that of a small halogen flashlight.)

- How often should the unit be checked for calibration, or recalibration—one or more times a day, a week, a month, or other?

- Does the unit's calibration shift with a lamp change (other than "0")?

- Does the unit have the capability of being linked to a computer, and how complete are the instructions for making this link?

- How accurately does the unit (if wide-band) read a standard reference material like T-Ref? Is the unit's accuracy in reading T-Ref specified by the manufacturer? Also, is the unit's repeatability specified by the manufacturer?

- How durable is the unit if dropped from table-top height (about 29")?

- How easy is it to change the unit's lamp?

- What is the manufacturer's or dealer's repair/loaner policy? Who has used it?

- What is the unit's aperture size? Make sure the aperture size is large enough for accurate print characteristic evaluation and small enough to measure your color bar and other print targets. **Note:** If you are measuring images printed using coarse screens with a densitometer having a small aperture, readings can become less accurate and less repeatable (see Appendix C).

Consider that the answers to most of the above questions are either right or wrong only when compared to other units for performance and cost, and only after a serious evaluation of the features you need. However, you should take none of these points lightly.

How sensitive are densitometers to surface irregularities?

Densitometers conforming to ANSI/ISO standards are quite sensitive to the reading surface. Although these standards call for densitometric readings to be made on a hard, flat surface, users will sometimes make measurements on soft, pliant surfaces, such as a stack of printed sheets. To discover the impact that an uneven surface can have on the stability of a unit's "Z" or Up-and-Down position, use the following procedure:

1. Zero the unit on a hard, flat surface.

2. Read a color bar or other target that is flat on this hard surface and record the measured values.

3. Read the same color bar or target with the densitometer evenly shimmed up with a file card or other piece of similar stock, and record this second set of measured values.

4. Read the same color bar or target with the densitometer only partially shimmed up with a file card or other piece of similar stock, and record this third set of measured values (see Diagram 6).

DIAGRAM 6: A
densitometer conforming to ANSI/ISO standards will be sensitive to small changes in surface contact.

Flat on sample

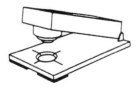
Elevated with shim

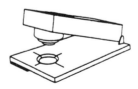
Tipped with shim

Notably, *an instrument whose measurements show little difference when the unit is fully or partially shimmed is probably not very sensitive to surface irregularities, and probably does not have the 5-degree limitation on angles specified by the ANSI/ISO standard.* Lack of conformance to this standard suggests that the unit is not well focused on the surface, which can contribute substantially to the overall variation of a densitometer's readings. One result of using units not in conformance with the standard is that users will not know with confidence when a small, but real, change in a printed density or other attribute occurs.

VI.   Improvements in Densitometry

What improvements have occurred in densitometry?

In numerous ways densitometers have improved significantly over the last five to ten years. According to a Graphic Communications Association survey of agencies, publishers, separators, printers, suppliers, and manufacturers, these improvements include the following:

- A reflection standard ISO 5/3 (Status T) now exists to define the spectral product of reflection densitometers. Manufacturers now have a standard to aim for, and users have a mathematical definition from which the absolute density of non-neutral samples can be determined via traceability through National Institute of Standards and Technology (NIST) certified tiles.

- Electronic precision and control of components—including better sensors, light sources, and filter arrangements—give today's densitometers more consistent results. Photomultipliers have been replaced with photodiodes and silicon planar sensors, providing a linearity and stability that were almost impossible to achieve with photomultipliers. Filter sandwiches are hermetically sealed. FET amplifiers have improved by at least one order of magnitude in stability.

- Digital electronics have also improved instrument accuracy, reliability, and consistency, shortening or eliminating the need for a warm-up period and, even more important, reading variations caused by heat, since line voltage fluctuation does not bother newer solid-state models. Instruments are much more stable and operate with less reading noise (some densitometers have a x10 expansion so as to read three numbers after the decimal). New instruments hold their calibration over a longer time period so that the need for frequent recalibration throughout a working shift has been reduced. Instruments now require less maintenance and are more robust and less prone to failure.

- Small tungsten-halogen lamps suitable for reflection densitometers have become available, providing a brighter and more stable light source.

- Digital read-outs are more precisely read and understood. The read-out areas themselves have been redesigned to minimize eye fatigue. Digital readouts provide resolution that can make use of the extra linearity and stability possible with new equipment.

- The design of the instruments makes the measuring head more comfortable, with the power cord coming from the middle of the back so that it can be operated easily using right or left hands.

- Calibration has always been the biggest problem in densitometry. The ease of calibration with newer equipment—it is now user-friendly—is among the most important advances in densitometry.

- Microprocessor-based enhancements have expanded the range and analytical functions that densitometers are capable of performing. These enhancements include direct reading of dot area percentages, trap, hue error, grayness, and efficiency. These enhancements have allowed for programmable features, memory capabilities, and the ability to interface with computers for analytical purposes, providing information quickly and accurately and making instruments easier to operate. Instruments can be zeroed easily, and changing modes to read density and dot gain is easier.

- Instruments are more portable.

- Data communication capabilities are now possible, providing direct data input into computerized systems to allow film and proof evaluation using a database of previous readings. Most densitometers can also be equipped with printers.

- Smaller and wider apertures are optional choices.

- Combination units are available that merge reflection and transmission capabilities.

- The ability to measure through visual, red, green, and blue filters simultaneously is now available.

This survey also revealed that the introduction of narrow-band densitometry presents no particular improvement. Narrow-band instruments produced by different manufacturers cannot provide effective inter-instrument agreement until there is effective standardization of a narrow-band response.

VII. Standards and Verifying Conformance to Standards

About the ANSI/ISO standard

Approved in 1984, ANSI Standard PH2.18/ISO Standard 5/3 for *Density Measurements—Spectral Conditions* defines a set of numbers indicating the standard spectral response for:

- **Status A,** which defines the response function for the measurement of photographic products such as photographic prints, 35mm slides, and transparencies that are intended for visual display using direct or projection methods.

- **Status M,** which defines the response function for the measurement of preprint films, which are camera films that are to be printed on photographic paper but are not to be viewed directly, such as internegative films used to create color prints.

- **Status T,** which defines the response function for the measurement of off-press proofs, press proofs, and press sheets, and other graphic arts materials being measured using wide-band equipment.

The only practical way that densitometer manufacturers and users in the printing and publishing arena can determine whether or not their Status T equipment is conforming to the standard is through the use of a printed paper reference that is fully traceable through the National Institute of Standards and Technology (NIST). The GCA T-Ref™ fits that specification (see Appendix A).

At the time of this third printing (May, 1992) ANSI Subcommittee IT2-28 is completing a narrow-band response standard, a standard for equipment using the 47B wide-band approach, and other relevant standards pertaining to graphic arts and photographic densitometry, colorimetry, and spectrophotometry.

About the DIN standard

A densitometry standard has also been developed by the German standards-making organization, Deutsches Institut fur Normung e.V. (DIN). DIN Standard 16536, entitled *Color Density Measurements On Prints: Requirements on Measuring Apparatus for Reflection Densitometers,* includes three parts, covering terms, requirements for color density measurement equipment, and measurement procedures. In contrast to the American (ANSI) and international (ISO) standard, the DIN document does not contain a response standard.

Summarizing a translation of the DIN standard, the document specifies measurement geometry, the type of light for the sample illumination, radiation receivers, measurement filters, how measurement values should appear to the observer, the size of the measurement field, equipment linearity, and the measurement calibration method.

A summary of the DIN standard indicates:

- **Measurement Geometry**
  45°/0° or 0°/45°, ± 5°, with an annular ring system recommended when measuring substrates such as paper.

- **Type of Light for the Sample Illumination**
  Illuminant A (2854K ± 100K)

- **Radiation Receivers**
  Photocell receivers must be "sufficiently sensitive. . . between 380nm and 720nm. A sensitivity present in the near infra-red (above 720nm) must be eliminated by means of an infra-red blocking filter."

- **Measurement Filters**
  The filters can be located ahead of or after the sample

in the illuminating path. (Note: ANSI/ISO has rejected this optional approach to the location of the filters because of the potential measurement differences that will result from this variable. The impact that this difference might have is referred to in the DIN document as "inconsequential.")

Filter color density refers to measurements made with a "spectral weighting" approach that uses an absorption filter, an example of which would be the Wratten filter. This weighting has the intended function of serving as a spectral product like the ANSI status response functions, but, unlike ANSI, specifies only the filter response and not the overall instrument response. For measurement of filter color density, DIN specifies:

- **To measure yellow**, use a blue filter peaking at 430 ± 5nm and having maximum band-width of 40nm at 50% transmittance and 80nm at 10% transmittance (see Diagram 7A).

- **To measure magenta,** use a green filter peaking at 530 ± 5nm and having a maximum band-width of 60nm at 50% transmittance and 100nm at 10% transmittance (see Diagram 7B).

- **To measure cyan,** use a red filter peaking at 620 ± 5nm and, with infra-red suppression in place, having a maximum band-width of 50nm at 50% transmittance and 100nm at 10% transmittance (see Diagram 7C).

Filters must also read within ± 0.1 density units of the spectral color density defined by DIN.

Conventional use of the term "spectral color density" refers to monochromatic measurement, which is the evaluation of colorant transmittance, reflectance, or density at one wavelength. Values obtained using a spectrophotometer are considered in the DIN standard to be monochromatic or spectral. For measurement of spectral color density, DIN specifies:

- **To measure yellow,** use a blue filter peaking at 430 ± 5nm and having a maximum band-width of 20nm at 50% transmittance and 30nm at 10% transmittance (see Diagram 7A).

- **To measure magenta,** use a green filter peaking at 530 ± 5nm and having a maximum band-width of 20nm at 50% transmittance and 30nm at 10% transmittance (see Diagram 7B).

- **To measure cyan,** use a red filter peaking at 620 ± 5nm and having a maximum band-width of 20nm at 50% transmittance and 30nm at 10% transmittance (see Diagram 7C).

DIAGRAM 7A: *Spectral reflectance of a yellow solid print and transmission of DIN filters to measure filter color density and spectral color density (infra-red trimming filters included).*

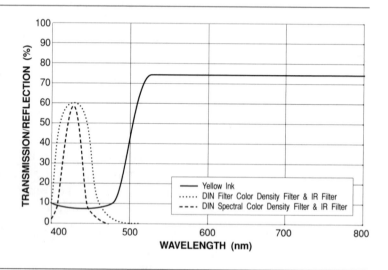

DIAGRAM 7B: *Spectral reflectance of a magenta solid print and transmission of DIN filters to measure filter color density and spectral color density (infra-red trimming filters included).*

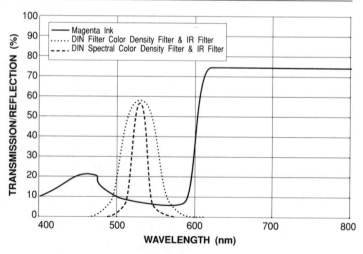

The DIN document does not give definite guidelines with regards to design of the equipment that should be employed when measuring spectral color density. This is key because even spectrophotometers are, in practice, not able to measure in a precisely monochromatic manner. Thus, two spectrophotometers that are not carefully set up and calibrated can provide different "monochromatic" values for a single colorant. Finally, in order to assure that a densitometer is perceiving light in a specified manner, the DIN standard is not adequate because the rejection of at least one density range,

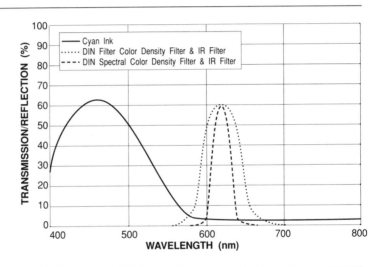

DIAGRAM 7C: *Spectral reflectance of a cyan solid print and transmission of DIN filters to measure filter color density and spectral color density (infra-red trimming filters included).*

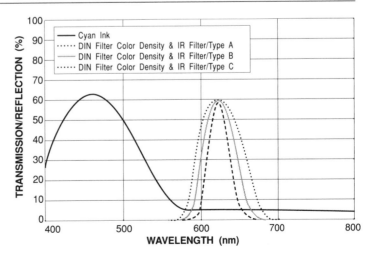

DIAGRAM 7D: *Filters with the curves indicated all meet DIN standards, but will not yield identical readings of cyan ink.*

called a "decade," beyond the highest density to be measured must be indicated. Thus, for printing and publishing applications using reflection densitometry, it is important to specify how the unit will perceive a density of 3.0.

- **Appearance of Measurement Values to Observers**
Equipment should display densities in direct numeric terms and to two places, e.g., 1.61, 0.53, 0.95.

- **Aperture Size**
The densitometer's aperture diameter should be 3.5mm, although 3mm is specified as the absolute lower limit. These diameters were selected based on screen rulings of 60 lines per centimeter (150 lines per inch). DIN notes that for coarser screen rulings the size of the aperture must be tested. DIN notes also that the measurement area must not be affected by areas surrounding the spot being measured, including paper white. DIN then suggests a method of determining the smallest area size that can be measured before the surrounding area begins to affect readings. In contrast, ANSI PH2.17 specifies the minimum aperture to be 0.5mm. The ANSI standard also specifies uniformity of illumination and sensitivity across the measurement aperture and states that non-uniformity of these functions leads to variation when measuring non-uniform samples; paper is a non-uniform sample. The DIN document does not address these factors.

- **Equipment Linearity**
Equipment must be linear to within ± 0.02 density units from a perfectly calibrated gray-scale target. If the gray-scale target varies ± 0.02 density units, the user can expect the total variation of the equipment readings to range within ± 0.04 density units from the linear. The densitometer must maintain linearity with the addition of filters.

- **Measurement Calibration**
Equipment should be calibrated using an ideal white as the reference, defined by other DIN standards. The document notes that in the case of a correct calibration of the instrument or a correction in agree-

ment with the measurement value, a graph of the unit's linearity should pass through the null point of the graph.

Another measurement method covered by ANSI—but not by DIN—concerns over-illumination. ANSI PH2.17 specifically requires that the illumination area of the target exceed the measurement area by 2mm. This over-illumination reduces errors in readings caused by light being reflected *beneath* the surface of the substrate as well as *off* the substrate's surface. This effect can be substantial in the printing and publishing process because the rough surface, lack of a hard reflector coating (such as those common in photographic papers), and other physical properties of printing paper allow a lot of light to reflect from beneath the surface.

Several points should be considered with regards to the DIN standard:

- It specifies a geometry and illuminant identical to that described by the ANSI/ISO standard.

- Although the standard specifies components—including selected transmittance values for the specified filters—it is not a response standard, which means that it does not specify precisely what spectral (color) response the densitometer should have. To users, the allowable variation in DIN can allow manufacturers to conform to the DIN standard and yet produce densitometers that will not allow communication of density numbers. This allowed-for variation also means that a check device that has any value in determining DIN conformance—such as the T-Ref has for determining ANSI/ISO conformance—cannot be developed. On the other hand, it is possible that a densitometer made to DIN specifications could, in fact, conform to the ANSI/ISO Status T standard. The way to check whether this is the case would be by using T-Ref.

- It allows manufacturers an option regarding the type of filter to be included in the unit, with one set being wide-band, the other narrow-band. One result: the user must know which DIN filter set is being used in the equipment (see Diagram 7D).

APPENDICES

Appendix A:
The Process of Using the
GCA T-Ref™

Status T is the ANSI/ISO standard for wideband densitometer response for measuring printed sheets and off-press proofs. T-Ref allows the printing and publishing industry to check whether densitometric equipment is calibrated by the manufacturer to Status T. More than one thousand are in circulation today, with an ever expanding number of users finding value in being able to communicate density numbers accurately and with assurance that they are speaking the same language using Status T.

Over time, T-Ref users have posed many thoughtful questions on the application of T-Ref. Here are some answers; let the Graphic Communications Association know if you have other points in need of clarification.

What is the GCA T-Ref ™?

The GCA T-Ref™ is a laminated paper reference printed with SWOP inks for use by agencies, publishers, separators, printers, and suppliers to determine whether their wide-band reflection densitometer conforms to the ANSI/ISO standard. By assuring conformance to the standard, T-Ref enables users to compare numbers from different densitometers with certainty.

Although T-Ref might have been produced using ceramic tiles or other non-graphic arts materials, GCA's Densitometry Task Force purposely chose an ink-on-paper reference so that readings would correlate well with what operators would experience in a production setting. Earlier research by Task Force member Miles Southworth supported this decision: In his study undertaken for the Gravure Technical Association, now the Gravure Association of America, Southworth concluded that "any standard must be as much like ink-on-paper as possible for it to be valid."

How are T-Refs calibrated?

In his previously cited article, Southworth notes John Yule's research suggesting use of a spectrophotometer to calibrate reflection densitometer reference materials. Franc Grum, who before his untimely death led RIT Research Corporation and Munsell Color Lab efforts to calibrate T-Refs, was the first to attempt carrying Yule's proposal to a practical conclusion. After making needed adjustments to spectrophotometric equipment—advancing the state of the art of this measurement technology in the process—Roy Berns, who succeeded Grum as Director of the Munsell Color Lab, Milt Pearson of the RIT Research Corporation, and William Voglesong of PSI Associates were able to refine the calibration system to meet industry demands for consistency and accuracy. T-Refs are now calibrated at PSI Associates, a laboratory specializing in sensitometric calibration, consulting, and research.

How do T-Refs fit into a production setting?

The T-Ref is designed to let densitometer users know in a practical manner how close to true Status T wideband reflection their densitometer comes. The T-Ref consists of printed samples of cyan, magenta, yellow, and black ink patches, with eleven readings calibrated to the ANSI/ISO Status T response, with calibration traceable to the National Institute of Standards and Technology (formerly the National Bureau of Standards). Users check patches on their T-Ref with their densitometer and compare these readings with the T-Ref's calibrated values. If they do not agree, the user should review the data with the manufacturer for possible recalibration of the instrument. By using T-Ref, the user can be assured at all times that the densitometer is still reading as accurately as possible.

How close should my densitometer's readings be to the T-Ref values?

Very. This cannot, however, be exactly defined because it depends on the agreement reached between the user and the densitometer manufacturer when the densitometer was purchased. The user should work this point out before testing with T-Ref.

How close are T-Ref's readings to the ANSI/ISO standard?

Within less than 1% of the ANSI specified response, traceable to the National Institute of Standards and Technology (NIST), and warranteed by GCA and PSI Associates. The total uncertainty of the T-Ref numbers—the difference between what the ANSI/ISO standard would specify as the density for a T-Ref spot and what the T-Ref's computer-generated label states is the density for that spot—is the greater of 0.02 density units or 2%.

How does T-Ref differ from the SWOP Hi-Lo reference?

The T-Ref is an instrument reference; the SWOP Hi-Lo is a proofing control reference. The T-Ref is used periodically to check the manufacturer's calibration of a wide-band Status T densitometer so that users may use the numerical reading of the densitometer to evaluate live work. The SWOP Hi-Lo Reference provides high and low density swatches within which the densities of proof materials should fall.

What does SWOP say about Status T and T-Ref?

Page 21 of the 1988 edition of *Recommended Specifications for Heatset Web-Offset Publications (SWOP)* states:

> Until 1984 there were no published standards for instrument responses at various wavelengths of light. In 1984 the American National Standards Institute published ANSI National Standard PH2.18 *Density Measurements—Spectral Conditions* defining standard densitometer performances. The portion of this standard that is relevant to the graphic arts is "Status T" densitometer response. With the acceptance of ANSI Standards for Status T Response for reflection densitometry, the availability of instruments conforming to this standard from most manufacturers and the availability of the GCA "T-Ref" for verifying such conformance, it is felt that numerical readings of one densitometer might well be able to be used as references for another in some applications.

How often do I use T-Ref?

Periodically, such as once a week depending on densitometer use, and at any time as a starting point when the user does not get the readings expected from live work.

Can T-Ref be used with any kind of work?

Yes.

Do I calibrate on T-Ref?

The answer depends upon the manufacturer's instructions regarding calibration. Here's why: for the end user to read Status T values on printed and proofed products, the equipment manufacturer must first build Status T response into the equipment. This calls for using the most appropriate filters, lens, optics, computer programming, and other elements inside the unit. The unit can drift away from the calibration built into the unit at the factory. To provide a means for allowing the user to adjust equipment when this occurs, manufacturers ask users to calibrate it and provide instructions on how this can be accomplished.

Calibration of equipment is sometimes handled using a manufacturer-supplied calibration plaque. Other manufacturers specify use of the black and white patches on T-Ref as calibration aim points. From a metrology (science of color measurement) standpoint, both approaches are acceptable. T-Ref's value as a third-party check on the unit's response is undiminished when calibrating to the black and white T-Ref patches; the color patches are used to verify equipment conformance to Status T response. **In all cases, however, follow the manufacturer's instructions when recalibrating since each company has developed procedures specifically for its equipment.** (Editor's note: in more technical terms, calibrating on white and black, whether on T-Ref patches or a separate target, serves to establish the photometric level of a densitometer. Having established this level, a densitometer with good spectral response will read the color patches on a T-Ref within the stated tolerances without further adjustment. Instruments may be designed for minor adjustments of individual colors.)

What happens if I calibrate equipment on T-Ref color patches?

It depends on the densitometer: those built with Status T response could be calibrated to T-Ref (black, white, and colors) and still read printed materials and provide Status T values. There are two important reasons why calibrating on the T-Ref color patches is **not a good idea:**

1. T-Ref is a third-party **check** on the densitometer's built-in Status T response. Brand new densitometers—even if the label says the unit conforms to the Standard—may not accurately read Status T response; older equipment may suffer from a hardware or software failure that alters the equipment's capabilities to read printing with Status T. Even without a failure of a densitometer component, all equipment varies and all densitometer components vary. For all of these reasons, users need T-Ref as a guaranteed reference with which to **check** equipment. Calibrating to T-Ref color patches instead of following the manufacturer's calibration instructions prevents the T-Ref from serving as this neutral check.

2. Calibrating to T-Ref color patches will suppress, to some extent, the equipment's built-in response. Equipment manufacturers design and build their equipment to be calibrated in a specific manner. Not following manufacturer procedures can lead to unwanted measurement variation. Specifically, calibrating to T-Ref color patches may not affect accurate Status T readings when ink densities are in a median range of approximately .80 to 1.70, but the equipment's now-altered curves may not accurately measure printing at higher or lower densities.

Why do I need to replace T-Ref?

Some deterioration in these materials is possible. To make T-Ref as similar as possible to the ink-on-paper world that we measure every day, GCA's Densitometry Task Force decided that T-Ref should be printed with SWOP inks on a white paper whose shade fell in the middle of white paper hues in common use. A comprehensive study undertaken by GCA in concert with Bill Voglesong of PSI Associates, who Chairs the ANSI

IT2 Committee that oversaw development of Status T, indicates that with normal wear and tear, the Status T densities generated for T-Ref are valid for at least a year, and possibly longer. To prevent any miscommunication, Densitometry Task Force members agreed to be conservative and to state that the validity of the readings could not be guaranteed for any time period longer than 12 months from date of the T-Ref's purchase. It's thus recommended that the user replace T-Ref each year; to make this decision process easier, GCA offers a discount on replacement T-Refs with the return of the current unit.

What's a good way to use T-Ref in my operation?

Here's one suggested approach; others exist that you could use.

- **First**, identify one person in each plant—or in the entire company—who is responsible for calibrating your densitometers and checking them with T-Ref. As an alternative, identify one person per shift or in each department who will take on this responsibility. **Reason**: Not all plant people will have the time or knowledge to calibrate and check your equipment accurately. Also, each person—even if they are properly trained and have an accurate understanding of how to calibrate the equipment and use T-Ref—will undertake this task in a slightly different manner. Having one person responsible for these functions minimizes this variation.

- **Second**, even if you have more than one densitometer, calibration plaque, or T-Ref, select one densitometer, one calibration plaque, and one T-Ref and use this set as masters. **Reason:** Equipment, calibration plaques, and even T-Refs all vary to some degree, however small, and by using just one of each as masters your measurement calibration process eliminates another variation source.

- **Third**, ask the specified measurement person to calibrate the master densitometer according to manufacturer instructions with the master calibra-

tion plaque and check it with the master T-Ref. **The designated measurement person should then use only the master calibration plaque and master T-Ref to calibrate and check all equipment in the operation.**
Reason: Undertaking this approach will eliminate the greatest amount of variation possible created by differences in calibration plaques and operators.

- **As an option:** This measurement person could use the master densitometer to read all other calibration plaques in the plant and note the densities on the plaques if they are different than the densities already printed on the plaques.

Using this approach you will minimize the variation inherent in any measurement system, whether that system be densitometric, colorimetric, or spectrophotometric. In this case it means your densitometers, if they are 1) built by the manufacturer to have Status T response, 2) calibrated by one person according to manufacturer's instructions, and 3) checked with a single T-Ref, should read within ±0.02 (some may creep into a ± 0.03 range) density of your densitometer population. Industry members have suggested, in fact, that expected variation among a group of densitometers is typically not greater than ±0.02.

Can variation among Status T densitometers be greater than ± 0.02?

Yes, even if the equipment conforms to Status T response a bigger variation is statistically possible, though not likely. Coupled with the ± 0.02 density unit variation that most densitometer manufacturers guarantee, your Status T equipment could, in a worst-case scenario, read a T-Ref patch with a density of 1.65 as being a value somewhere between 1.60 to 1.70 and still be in conformance with the standard. Here's why: 2% of 1.65 is greater than ± 0.02 density and thus is the figure to use in determining the T-Ref variation. This means that for no reason other than the variation inherent in the T-Ref calibration process, a properly calibrated Status T densitometer will measure this patch as being somewhere between 1.62 and 1.68. Adding the ± 0.02 densitometer variation extends the potential window of variability to 1.60 to 1.70.

This seems like a large window, but keep in mind that it's not likely that properly calibrated equipment will see this type of variation, although in instances it is possible. The ± 0.05 value is a worst-case situation: properly calibrated Status T equipment typically reads T-Ref values within a ± 0.02, or perhaps an ± 0.03, window.

Is improvement possible?

Can we narrow this ± 0.05 worst-case variation? Most of the variation seen in this measurement process is caused by the operator and methods of using the densitometer, not by the unit itself or T-Ref (for more detailed information on analyzing this variation, consult Appendix C). Using the procedure outlined previously will help minimize variation. In parallel, forward thinking densitometer manufacturers are working to reduce variation contributed by their equipment. PSI Associates, in concert with GCA's Densitometry Task Force, is also constantly evaluating potential refinements to the ANSI/ISO standards as well as the T-Ref production and calibration process to narrow the minimal variation contributed by T-Ref to this process.

What is "total uncertainty"?

The total uncertainty of the T-Ref numbers is the difference between what the ANSI/ISO standard would specify as the density for a T-Ref spot and what the T-Ref's computer-generated label states is the density for that spot. PSI Associates, working in concert with GCA's Densitometry Subcommittee, the Munsell Color Science Laboratory, the National Institute of Standards and Technology, and other metrology labs, has achieved a precision and accuracy of T-Ref readings so that they are no more than 0.02 density units or 2%—whichever is greater—from the true standard defined in PH2.18.

In technical terms, total uncertainty is the variation of the individual test (precision) and the spread of other standards labs (accuracy). Thus:

The *precision* with which PSI Associates makes T-Ref readings means how repeatable they are in making accurate readings. Precision measures T-Ref calibration repeatability. The deviation of 0.003 noted on the T-Ref label indicates the precision with which T-Ref calibrations are made.

The *accuracy* with which PSI Associates makes T-Ref readings means how closely they are able to measure T-Ref ink spots and calculate mathematically derived Status T values using real-world equipment and materials (Remember that everything varies, including each hardware and software component of the spectrophotometer used to calibrate the T-Refs as well as other elements, such as the NIST tiles used to calibrate the spectrophotometer.). Accuracy encompasses average variance from other standards laboratories as tested by a Measurement Acceptance Program (MAP), which conducts round-robin measurements of NIST tiles and analyses of these measurements.

The *total uncertainty* of the T-Ref values—0.02 density units or 2%—is the sum of the variation in the precision and accuracy in making T-Ref readings.

In practical terms, when a T-Ref indicates a value of 1.65 for the black spot measured with the visual filter, the true, mathematically derived Status T value is somewhere between 1.62 and 1.68 (since the $\pm$ 2% of 1.65 is greater than $\pm$ 0.02 density units), although we cannot exactly say where because of the variation in materials and measurements inherent in the calibration system. In comparison, most densitometer manufacturers guarantee that their units' readings alone have a total uncertainty of $\pm$ 0.02, meaning that a given densitometer reading of 1.65 could fall within a range of 1.63 to 1.67 without any reason other than the inherent variation of the measurement system in that unit.

This outstanding achievement of a total uncertainty of 0.02 or 2% for the T-Ref means that T-Ref values, regardless of the natural variation of the calibration system, are very close to true Status T and as such are excellent indicators of whether a given densitometer was calibrated to Status T response or has drifted away from maintaining this response.

Summary: it's good

We've made enormous strides in measurement technology to be able to accomplish this fairly tight window and make these statements regarding densitometer performance. Further metrology research is now being undertaken to refine this variation still further. Keep in mind as well that colorimeters and spectrophotometers being widely touted as "solutions" to this variation have just as much, if not more, variation inherent in their use.

Appendix B:
Standard Lighting

Why we use standard lighting

ANSI PH2.30-1989: for Graphic Arts Photography—Color Prints, Transparencies, and Photomechanical Reproductions—Viewing Conditions defines the requirements for standard lighting conditions in the printing and publishing industry. The standard addresses the light's minimum color rendering index (also called CRI), luminance (candelas per square meter), illuminance (lux), uniformity (percent of nominal), spectral power distribution, correlated color temperature (Kelvins), and surround conditions.

Even though consumers usually look at printing in room light or daylight that can be brighter or darker, and perhaps also "warmer" or "cooler," than a standard light source, print production professionals use the ANSI standard because:

1. A 5000 Kelvin standard light source falls in the middle of the range of light that humans are exposed to, and is a color-balanced white light that does not emphasize one color more than another.

2. Color experts in the print production process must visually evaluate, analyze, and compare the color appearance, density, and surface characteristics of a wide range of color materials, including transparencies and other original artwork, off-press proofs, and press sheets. Commercial lighting is not suitable for these viewing requirements because it is color deficient—it is not color-balanced. Natural daylight is balanced, but it is constantly changing and, thus, not suitable.

"Standard" lighting may not be

A 5000 Kelvin light source used for color image evaluation gradually changes in color balance and total light output as the fluorescent lamp's phosphors degrade with use. How do users check whether a lighting booth remains in conformance with the ANSI standard? Experts suggest the following ways:

1. Install a lamp-use monitoring meter, replacing lamps within the intervals specified by the viewing system manufacturer.

2. Use photometric measurement to determine the light source output. These meters provide readouts in footcandles or footlamberts and can indicate whether the viewing system is within specification, e.g., 204 ± 44 footcandles for reflection viewers, and 408 ± 88 footlamberts for transparency viewers.

An option: use radiometric-colorimetric measurement. Unfortunately, inexpensive, hand-held color meters will not indicate the absolute colorimetric or accurate, correlated color temperature values for the viewing system. For such benchmark measurements, a laboratory-type spectroradiometer is required. Devices also exist that are marketed as a quick way to determine whether a lighting system is emitting the proper light. Consisting of two ink patches, these devices are based on the concept of metamerism, whereby the inks on the card appear identical under one light source but different under another one.*

---

* **Caution**: Experts indicate that, although these devices are very effective for dramatizing the fact that incandescent or cool white fluorescent lighting does not have good color rendering properties, they are **not** effective in analyzing a viewing system. The ink metamers used on these cards will appear very close in color appearance under a fairly wide range of correlated color temperatures, ranging from about 4500 Kelvin through 7500 Kelvin, assuming the lights have a good color rendering index. Also, **the single greatest cause of viewing deficiencies in our industry is improper fluorescent lamp maintenance, and these metameric devices will not diagnose this anomaly.** When considering this, it is important to note that the ANSI lighting standard specifies five conditions: light quality, light intensity, evenness of illumination, geometry of illumination/viewing, and surround conditions. Aimpoints for these five conditions must be met for adherence to the standard. The cards using metameric inks only address the first condition, and that not very precisely.

Appendix C:
Standard Methodologies for Determining Densitometry Capability

To understand variation caused by the densitometer and operator together

Making measurements with a densitometer is a process. This process includes inputs from the instrument, operator, procedures, reference standards, and data recording and reporting. Each of these inputs has an inherent variation, and each contribution of variation accumulates to affect the final densitometric measurement. **Densitometry** includes all of these inputs and must take into account their contribution of variation. **Densitometers** are only one component of the system. This becomes important if a manufacturer claims that equipment is repeatable within a ± 0.015, or similar, tolerance. Though a true statement, confusion can be created if it is not specified whether this range encompasses variation contributed by any component of the densitometry measuring system other than the densitometer itself.

Users need to understand how accurate and precise both their equipment and their measuring process is, and what methods they should use to evaluate variation inherent in the total densitometer measurement system.* Recall first that a densitometer measurement system has three major components: 1) an illumination system; 2) a collection and measurement system; and 3) a signal processing system. Remember also that

---

* The ANSI Committee IT2-28 responsible for developing standards pertaining to graphic arts densitometry is currently at work on confirming the standard densitometer capability studies, such as those described in this appendix, that will allow users to evaluate different contributors to the overall variation of the densitometer measurement system. Based on material originally presented by IT2-28 Chair William Voglesong of PSI Associates at a GCA Densitometry Tutorial, these studies examine variation in measurement processes such as the operator-and-densitometer measurement system, the densitometer portion of this measurement system, and the photometric portion of the densitometer. They employ statistical analysis techniques considered to be a standard part of statistical process control (SPC) practice. For additional information about this text, statistical process control charts, and other helpful materials about SPC, contact GCA at 100 Daingerfield Road, Alexandria, VA 22314-2888.

everything varies to some degree. A measurement capability study will alert you only to variation that is greater than your application demands. It will not indicate that the unit being checked is performing without variation.

Although these methodologies encompass a range of descriptions pertaining to densitometer components and measurement methods, users may want to undertake other efforts as their process needs require and experience broadens. For example, these methods can also be applied to understanding colorimeter variation, allowing users to study differences between densitometry and colorimetry measurement systems.

1. Calibrate your densitometer, and check it using T-Ref, as you would normally in a production setting.

2. Using the T-Ref, which has cyan, magenta, yellow, black, and white patches identified with calibrated densities, or a manufacturer's supplied target, take three readings of each patch each hour for at least 20 hours without turning the densitometer off. You do not need to take the readings for 20 sequential hours; readings can be taken over two or three days as long as the densitometer is not turned off (battery-powered portable units go into a dormant or "sleep" mode between readings; if the charger is activated, these units are essentially "on.") Turning the equipment off could create a new set of variables. The readings can be taken in any sequence. Suggested is white-black-cyan-magenta-yellow, repeated three times.

3. Analyze this data using a GCA Statistical Process Control Chart (an X-bar/R chart). Using your twenty subgroups of three readings (for good statistical insight, it is recommended that a minimum of twenty subgroups be used for a statistical analysis like this), follow the instructions on the chart to plot subgroup averages and calculate upper and lower control limits and the range. Do an X-bar/R chart for each patch—white, black, cyan, magenta, and yellow.

4. Results will indicate the total uncertainty of the measurement "system" that includes the operator and densitometer being evaluated. For example, the magenta X-bar/R chart could have an average (X-double bar) of 1.38 with an upper control limit of 1.44 and a lower control limit of 1.32, with all plotted points falling between these upper and lower limits. This suggests that the operator-plus-densitometer measurement system is statistically stable for magenta. Note that this example range of readings is broader than the tolerance manufacturers claim for their equipment. This is expected, and is a result of the **total, cumulative** variation of operator-plus-densitometer measurement system as opposed to the variation of the densitometer alone.

You thus know what stable variation is occurring with the operator-plus-densitometer, providing the understanding that the actual density of a patch read by that operator and unit could actually be anywhere in a ±.06 range around that point. If the operator uses that unit to measure a press sheet, and the magenta is read as 1.28, the density could actually be anywhere between 1.22 and 1.34.

This technique allows you to compare:

- The variation of one densitometer to another, so long as the same operator uses both units when making measurements and the same targets are used.

- The variation of one operator to another, so long as the same densitometer is used by both operators and the same targets are used.

To understand variation caused by the densitometer alone

Positioning of the densitometer on a surface—including where the operator places the unit and how he or she brings the unit's head down to make a measurement—allows variation. This study permits understanding how great this variation is by studying what variation is contributed by the placement and the up-and-down, or "Z," movement of the densitometer alone.

1. Calibrate your densitometer, and check it using T-Ref, as you would normally in a production setting.

2. Cut out a target from a press sheet or proof, and tape it securely over the hole in the densitometer shoe. Check to make sure that when the densitometer head is brought down, the patch covers this hole completely.

3. Ask the operator to bring the head down once every two minutes until at least 20 subgroups of three readings (60 readings) are taken.

4. Repeat this for each patch—white, black, cyan, magenta, and yellow—and analyze each using a GCA Statistical Process Analysis chart.

5. Results will indicate the total uncertainty of all aspects of the densitometer portion of the measurement system.

This technique allows you to compare:

- The variation of the densitometer alone.

- The variation of one densitometer to another.

To understand variation caused by the environment

Studies of glossmeters equipment used in the paper industry indicate that room temperature can have a dramatic impact on equipment stability. Densitometers, which from an equipment design standpoint are similar to glossmeters, could be similarily affected. This study permits understanding how great this variation is by measuring what variation is contributed by room temperature.

1. Confirm statistically that the variation caused by the densitometer's reading head is stable.

2. Develop a mechanism that allows the operator to trigger the densitometer measurement via remote control, using a computer. This allows the unit's head to be in the "down" position at all times, removing any variation created by placement or up-and-down movement. This computer program can be developed internally or purchased. For infor-

mation on where to get a program that accomplishes this, contact GCA.

3. Set the temperature of the room where the unit is located at the coolest level it might get, or move it to a cool area. This low temperature should represent the coldest environment possible when the room is air-conditioned, or when the heat is low or turned off during winter.

4. With the room temperature low, ask the operator to use the computer to trigger the densitometer light source once every two minutes until at least 20 subgroups of three readings (60 readings) are taken.

5. Repeat this for each patch—white, black, cyan, magenta, and yellow—and analyze each using an X-bar/R chart.

6. Set the temperature of the room where the unit is located at the warmest level it might get, or move it to a warm area. This high temperature should represent the warmest environment possible when the room is heated, or when the air conditioner is low or turned off during summer.

7. With the room temperature high, ask the operator to use the computer to trigger the densitometer light source once every two minutes until at least 20 subgroups of three readings (60 readings) are taken.

8. Repeat this for each patch—white, black, cyan, magenta, and yellow—and analyze each using an X-bar/R chart.

9. Comparing the averages and ranges of the two sets of X-bar/R charts will indicate the impact, if any, that room temperature has on the densitometer's measurement system.

William Voglesong has suggested two ways to increase the temperature of the unit without changing the temperature of the entire room: 1) drape an electronic blanket over the unit and turn it to a temperature that is not excessive; 2) shine several bright lights, such as photographic studio lights, on the unit. In all cases, remember to record the temperature of the environment surrounding the unit.

To understand variation caused by the light in the surrounding environment

Although lighting in the area surrounding a densitometer usually does not affect modern units, the impact of this light on a unit, which if it gets inside the densitometer is called "stray light," can have an impact on measurement variation. To evaluate this factor:

1. Confirm statistically that the variation caused by the densitometer's reading head is stable, and that temperature changes do not affect the unit's stability.

2. Develop a mechanism that allows the operator to trigger the densitometer measurement via remote control, using a computer. This allows the unit's head to be in the "down" position at all times, removing any variation created by placement or up-and-down movement. This computer program can be developed internally or purchased.

3. Shine bright lights on the unit, or place it in a consistently brightly lighted area.

4. With the surrounding area brightly lighted, ask the operator to use the computer to trigger the densitometer light source once every two minutes until at least 20 subgroups of three readings (60 readings) are taken.

5. Repeat this for each patch—white, black, cyan, magenta, and yellow—and analyze each using an X-bar/R chart.

6. Turn off the lights, or place the unit in a dark area, or cover it.

7. With the surrounding area dark or dim, ask the operator to use the computer to trigger the densitometer light source once every two minutes until at least 20 subgroups of three readings (60 readings) are taken.

8. Repeat this for each patch—white, black, cyan, magenta, and yellow—and analyze each using an X-bar/R chart.

9. Comparing the averages and ranges of the two sets of X-bar/R charts will indicate whether light from the surrounding environment is straying into the densitometer and affecting measurements, as well as the degree of impact, if any, that surround lighting has on the densitometer's measurement system.

To understand variation caused by the densitometer's photometry

The physical act of bringing the densitometer reading head down can create variation. This study examines what total variation is created by the densitometer's illumination system, collection and measurement system, and signal processing system.

1. Confirm statistically that the variation caused by the densitometer's reading head is stable.

2. Develop a mechanism that allows the operator to trigger the densitometer measurement via remote control, using a computer. This allows the unit's head to be in the "down" position at all times, removing any variation created by placement or up-and-down movement. This computer program can be developed internally or purchased.

3. Ask the operator to use the computer to trigger the densitometer light source once every two minutes until at least 20 subgroups of three readings (60 readings) are taken.

4. Repeat this for each patch—white, black, cyan, magenta, and yellow—and analyze each using a GCA Statistical Process Analysis Chart.

5. Results will indicate the total uncertainty of the photometric portion of the densitometer's measurement system. This technique allows you to know the inherent capability of the light measuring portion of the unit, without adding variation from the operator or the positioning of the unit's head.

To understand when a unit should be recalibrated

Recalibrating a unit when it has not drifted can add variation rather than reduce it; not calibrating when the unit has drifted can also create problems. Experts suggest that recalibration is a function of use—the number of times readings are taken. To know the inherent capability of the light measuring portion of the unit to maintain calibration and not drift:

1. Use the mechanism that allows the operator to trigger the densitometer measurement via remote control, using a computer.

2. Ask the operator to use the computer to trigger the densitometer light source once an hour over four days, again without turning off the light source. Use the white patch for this portion of the capability study.

3. Repeat Step No. 2, only using the black patch.

4. Plot the data taken from the white and black patches using a GCA Statistical Process Analysis Chart on which is drawn the upper and lower control limits of the variation of the unit's photometric system.

5. When readings approach, touch, or exceed a control limit for the white or black patches, the chart will indicate the number of readings at which a unit should be recalibrated. This capability study is appropriate because drift is a function of the unit's measurement system, not the operator or movement of the reading head. You may find that even four days of data collected hourly indicates little need for recalibration. Conduct the effort for eight days, or longer, to determine when recalibration is needed.

This effort can also be used to compare two brands of densitometer, or to determine how well they maintain calibration.

To understand variation caused by aperture size

In reflection densitometry, aperture sizes vary from approximately 2mm to 4mm in diameter. Although densitometer aperture size is not a significant cause of variation when measuring a solid patch of paper or colorant, this component can have a **greater** impact when making measurements of halftone and tint areas. This happens because the aperture defines an area on the substrate that the unit illuminates and off which it collects reflected light. Smaller apertures define smaller areas on the printed surface, increasing the amount of reading variation that can take place because of slight changes in the placement of the unit on the sheet (see Diagram 8). Aperture size is, thus, a factor in the variation caused by the densitometer-plus-operator system.

To determine the extent of this variation, have on hand two densitometers manufactured by the same company and, preferably, as identical as possible in every way, except that one uses an aperture that is approximately 2mm, while the other uses a larger aperture, such as one measuring 3.5mm or 4.0mm. Then:

1. Calibrate both densitometers, and check them using T-Ref, as you would normally in a production setting. Make sure they both are conforming to Status T response.

2. Using a printed control bar with a screen ruling typically used in your operation (133 lines/inch, 150 lines/inch, 120 lines/inch, etc.), pick a 50% patch and measure it first with the large-aperture unit and then with the small-aperture unit. Record the large-aperture unit reading on one X-bar/R chart and the small-aperture unit reading on a second X-bar/R-chart.

3. Repeat this sequence, reading the same printed 50% patch and using the same operator, until a total of 120 readings are taken — 60 readings collected into subgroups of three readings will be recorded on the large-aperture X-bar/R chart, and 60 readings collected into subgroups of three will be recorded on

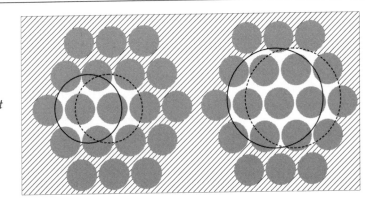

Diagram 8: *When taking measurements with a densitometer having a smaller aperture, minimal differences in placement of the unit can yield significant measurement variation.*

the small-aperture X-bar/R chart. **The operator should attempt to place the densitometer in slightly different spots on same the 50% patch.** The 120 readings can be taken one after the other or over some period of time. It is not, however, required that the readings be taken once every five minutes or at any other specified timed interval.

4. Analyze the two X-bar/R charts, comparing the averages and ranges of the data generated by the two units. Assuming the two units are identical (which will not be the case, although they should be very similar), and with the 50% patch and operator remaining the same, results will indicate how aperture size affects the variation contributed by the densitometer-plus-operator together.

Options helpful in exploring the extent to which aperture size affects a measurement system include:

• Measuring and analyzing the 25% and 75% patches, as well as the solids.

• Comparing statistical analyses of measured targets having different screen rulings or dot shapes.

• Undertaking this study using different types of paper, since the paper smoothness can also affect results (consider how ink laydown might look on a smooth cast-coated stock versus on rough newsprint).

Here's a quick option that, while not statistically rigorous, will still give insight into the effect that aperture size has on measuring any paper-ink-screen combination:

1. Measure one time the 50% (or other) tint using the small-aperture unit and record this measurement.

2. Measure ten times the same tint using the small-aperture unit. Record these measurements, and then average them. Record this averaged value. (Again, place the unit on the same tint in slightly different spots for a total of ten readings.)

3. If you've got a large-aperture unit, measure one time the same 50% (or other) tint and record this measurement.

4. Measure ten times the same tint using the large-aperture unit. Record these measurements, and then average them. Record this averaged value. (Again, place the unit on the same tint in slightly different spots for a total of ten readings.)

5. Compare the four resulting values. They will indicate:

—Whether a single measurement taken with the small-aperture unit is different from a single measurement using the large-aperture unit.

—The similiarity between several averaged small-aperture unit measurements and a single large-aperture reading.

—The difference between several averaged large-aperture unit measurements and a single large-aperture measurement.

Regarding aperture size versus screen ruling, it has been noted that as the screen ruling increases, the aperture size can decrease without any significant effect on measurement variability **caused by aperture size alone**. This is logical since higher screen rulings have more dots per inch, increasing the likelihood of a small-aperture unit measuring the same halftone pattern each time it is placed on the target.*

---

* While research by GATF reported at TAGA '89 indicates that, when measuring printed newsprint, larger apertures provide more consistent readings, in a production setting it is common to be forced to use a narrow—or half—color bar requiring measurement using a densitometer having a smaller aperture. As long as one is aware that smaller apertures have an impact on reading variation, it makes a great deal of sense to use narrow color bars and small-aperture equipment rather than to use nothing. Franz Sigg of the Rochester Institute of Technology T&E Center undertook a theoretical analysis using higher quality papers on the topic of reading variation contributed by the aperture diameter versus paper surface characteristics. His work suggests that for a 2mm aperture and an 85 line screen, a variation of ± 2% is possible, meaning that a 50% dot could appear to read anywhere from 48% to 52% because of the paper surface alone. Experts evaluating coarse screens on smooth paper report that in their testing they have only seen a contribution of ± 1% which means that a 50% dot could appear to read anywhere from 49% to 51% because of the paper surface alone. They note, however, that the ± 2% variation is probably realistic when taking into consideration all possible papers and equipment. What does this mean to the user? First, if practical, rougher substrates should be measured with larger aperture equipment; second, measurements of printing on rougher papers will see greater common-cause variation than measurements of printing on smoother papers.

Note that measurements made with smaller aperture units can still be more easily affected by factors such as paper smoothness/roughness, hickies, and ink lay-down variation. This is because the smaller aperture is defining and measuring a smaller portion of the printed or imaged substrate, making more important the impact these print phenomena have on the smaller area being measured.

Users who must use smaller aperture units to measure coarser screen rulings can still achieve the reading accuracy of large-aperture units. This is accomplished by taking several readings—at least three—of the control target (again, just read the patch—do not try to place the unit in exactly the same spot on the target) or other printed area and averaging these readings. Using this approach allows the smaller aperture to measure a larger printed area, simulating the amount of printed area that a unit with a larger aperture would measure in one reading.

Users might well develop additional methodologies using these approaches as outlines. The IT2 group developing them as standard methodologies welcomes suggestions both about these procedures as well as about new process capability evaluations. Comments in writing should be directed to GCA, 100 Daingerfield Road, Alexandria, VA 22314.

Appendix D:
Common Characteristics of Densitometers, Colorimeters, and Spectrophotometers

Some equipment manufacturers and users indicate that colorimeters and spectrophotometers offer more accurate and precise measurement values than do densitometers. Bill Voglesong of PSI Associates compiled a listing of how these three types of measurement technology compare and contrast, with insight into their real-world application and relative advantages and limitations.

All three share a common measurement approach and geometry

Spectrophotometers, colorimeters, and densitometers are all opto-electronic instruments with a common measurement approach. They all measure the reflectance or transmittance of off-press proofs, press proofs, press sheets, and other physical samples. These instruments all record both the light shown onto (incident on) the printed or imaged sample as well as the light either transmitted by or reflected off this sample. The sample is evaluated objectively by assuming that the light lost in this process is accounted for by the absorption. Writing this as an expression for reflection measurement we have:

1. $\Phi i = \Phi a + \Phi r$

Where:
Φi is the light flux incident
Φa is the light flux absorbed
Φr is the light flux returned

However, for actual samples and measuring instruments, all of the light is not accounted for in equation 1. Some light is scattered by the sample surface or by particles such as fibers or ink below the surface. Further, some light is spread within the sample itself. The irregular spreading of light in the sample is described by a point spread function ξ. Light spread in this manner may thus be written as Φ_ξ, while scattered light can be written as Φ_s.

Light falling on the sample is thus accounted for by a somewhat more complex expression:

2. $\quad \Phi_i = \Phi_a + \Phi_\xi + \Phi_s + \Phi_r$

Light reflected off the sample for measurement, Φ_r, is thus dependent upon the surface, surface scattering, and the surface point spread of the sample. The light perceived by the measurement device also depends upon the angles that the central rays make with the sample and on the cone angles of the illumination and collection beams. These parameters of measurement are called the system geometry, and they are specified in ANSI standards PH2.17-ISO 5/4 and PH2.19-ISO 5/2. The light loss by image spread, Φ_ξ, varies with the sample size and measurement aperture; PH2.17-ISO 5/4 includes definitions designed to minimize all these errors.

Design trade-offs required for building a practical instrument may cause some departure from the rigorous specification of these ANSI/ISO standards. This, of course, means some trade-off in reproducibility of measurement between instruments of varying manufacture or model. These trade-offs usually do not represent a major obstacle to quality measurement.

Spectral response: different slices of the same light

The fidelity of color measurement depends upon the total spectral response of the measuring instrument. Response is defined by the spectral product of the lamp power output, spectral modulation of optical components, and the photo-receiver sensitivity. ANSI standard PH2.18-ISO 5/3 defines this spectral product as:

3. $\quad \Pi_\lambda = \phi_\lambda t_\lambda s_\lambda$

Where:

Π_λ is the instrument spectral product
ϕ_λ is the lamp spectral power
t_λ is the spectral attenuation (lens, filter, grating, etc.)
s_λ is the spectral sensitivity of the receiver

For filter densitometers and colorimeters, t_λ is controlled by the selection of optical absorption filters. Addressing each measurement approach specifically:

1. For densitometers, the response, also called the status, is defined as Π_λ in ANSI PH2.18-ISO 5/3. The spectral responses encompassed by this document define densitometry accepted historically as standard by different groups of densitometer users. Specifically, Status A response has a relatively narrow pass-band used to provide a compromise between color separation and visual perception for photographic dyes used in products intended for direct viewing. Status M has a wider pass-band with overlapping values that emulate the overlapping spectral sensitivities of base material used for photographic prints. Status M is thus useful in predicting the performance of these pre-print materials. Status T documents the wide-band densitometer response commonly used in the North American graphic arts community, while Status E documents the densitometer response reported to be commonly used by the European graphic arts community. Status I documents the densitometer response referred to in North America as narrow-band.

2. For tristimulus colorimeters, the response equivalent of the Π_λ is defined by the CIE color mixture functions $\bar{x}, \bar{y}, \bar{z}$. The light source should be filtered to provide ϕ_λ for the viewing illuminant. Filter selection is sometimes compromised in practice to compensate for the illuminant specification.

3. For spectrophotometers, the spectral modulator t_λ may be a diffraction grating, a prism, or interference filters. The spectral sensitivity of the receiver, s_λ, is not critical because the pass-band $\Delta\lambda$ is small. For grating and prism spectrophotometers, the entrance

slit is imaged on the measurement plane. The width of the entrance slit determines the spectral purity and pass-band. The center of this image is at the nominal wavelength, λ, being measured. For interference filter spectrophotometers, the maximum filter transmittance determines λ while the side-band rejection of the filter determines the pass-band. Computer controlled spectrophotometers do not generally scan the full spectrum but instead sample at various intervals of $\Delta\lambda$.

Sampling with a $\Delta\lambda$ smaller than the pass-band does not increase spectral resolution but does provide data smoothing. Sampling with a $\Delta\lambda$ larger than the pass-band requires a more complex curve-smoothing algorithm. Increasing the pass-band provides a more stable measurement signal, but at the expense of resolution. Frequent sampling reduces the dependence on the smoothing algorithm, though at the expense of operating speed. **None of these trade-offs can be treated as best.** While pass-band and wavelength interval are often equal, this is not a necessary condition for good measurement. For reflection measurements, the most commonly used photographic dyes, printing inks, and paint pigments change so slowly across the spectrum that very narrow pass-bands and small spectral intervals are not required.

Signal within the spectral envelope Π_λ is an analog of the data desired, while signal outside Π_λ is noise and can be considered error. Good data needs a high signal-to-noise ratio, and thus little signal can be tolerated outside the intended measurement band. Spectrophotometers thus require better rejection than filter densitometers and narrow-band filter densitometers need better rejection than wide-band instruments.

Signal processing: one approach, different data formats

Measurements are the electrical analog of spectral components consisting of lamp energy, the return signal from the target, the spectral modulation of the instrument, and the response of the photocell. This value is given the symbol Πr, with the complete expression being:

4. $\Pi r = \int \phi_\lambda t_\lambda r_\lambda s_\lambda d\lambda$

The energy returned from a 100%, or 0 (zero) density reference, is the same as the energy incident on the target to be measured:

5. $\Pi i = \int \phi_\lambda t_\lambda s_\lambda d\lambda$

The ratio of return to incident energy weighted by the function 4, is called either reflectance or transmittance depending upon the nature of the target:

6. $R = \dfrac{\Pi r}{\Pi i}$

Densitometers display the output data as density, which by convention is calculated using a logarithmic function:

7. $D = -\log_{10} 1/R.$

For colorimetry, the three functions $\Pi_\lambda = \bar{x}, \bar{y}, \bar{z}$. The computation equivalent to equation 6 produces values for the tristimulus functions X, Y, Z. These tristimulus functions are the starting point for deriving any of a range of preferred colorimetric parameters such as L*,a*,b*.

Spectrophotometers provide for a data set $r_{\lambda i}$ for both incident and returned light. These data may be used to compute colorimetry by the formulae established by CIE or to compute density as specified in ANSI PH2.18-ISO 5/3.

Applications

Densitometry

Density values track chemical quantities because chemical concentrations, such as an ink layer, follow the Beer-Lambert law. Except for extra light returned by image surfaces, density varies in direct proportion to the amount of dye, ink, or pigment on the sample. As the concentration of the absorbing molecule increases, the sensitivity to change increases. For process control, density is thus sensitive, analytical, and predictive.

Colorimetry

Colorimetric values have been normalized to indicate color differences as perceived by the human eye. They lose sensitivity at high concentrations of the absorber. More specifically, color differences are measured in terms of linear difference in the three-dimensional color space. Color difference is expressed as ΔE. A value of ΔE is defined approximately as a just noticeable difference. Near white a $\Delta E \cong 1$ corresponds to $\Delta D \cong 0.005$. For dark samples, however, such as those where the density is near 1.4 or higher, the value $\Delta E \cong 1$ corresponds to $\Delta D \cong 0.03$. These data suggest that while colorimeters are more precise at alerting the user that a just-noticeable color change has occurred, densitometers are, in fact, more precise than colorimeters for controlling the process that yielded that color change.

Spectrophotometry

Spectrophotometric data is typically presented as either graphs of spectral data or tables of this information. While most complete in terms of its definition of color, these data are awkward to use in production settings since printing presses, off-press proof systems, and photographic lab processing methods rely on measuring and controlling shifts in primary hue densities and related reproduction attributes in order to control color. Spectrophotometric data may, however, be used to compute either densitometric or colorimetric values, making this device a powerful data collection tool.

Publications available from GCA. . .

Introduction to Densitometry: A User's Guide to Print Production Measurement Using Densitometry
This book is a non-technical introduction for press operators, plant managers, and print buyers to densitometers (the equipment) and densitometry (the process of using densitometers effectively). The book, which includes many helpful diagrams, is part of GCA's on-going program of expanding measurement understanding and application throughout the printing and publishing industry.

Introduction to Color Bars
This handbook answers key user and technical questions about color bar elements and color bar applications and analysis. It encompasses a range of topics including why color bars are key to color control, consistency terms important to color bar use, and methods critical to proper color bar application. Included are illustrations of commonly used color bars.

SPECTRUM Guide to Proofs and Pre-Proofs
A definition of proofs and pre-proofs, and an introduction to digital proofing and the requirements needed to fully use this promising new technology, highlights this document. Developed for SPECTRUM '89, the *Guide* includes a list of commercial proofing systems available. A section on pre-proof technology for the electronic front end also includes a list of suppliers of color printers, plotters and copiers.

GCA/GATF Proof Comparator II
The Proof Comparator is an easy and effective control strip used to evaluate off-press proofs. The solid, tint and, overprint patches, gray balance bars, star targets, dot gain scales, and exposure control patches allow the proof to be evaluated technically. The central pictorial, containing flesh tones and a wide range of colors, allows the proof to be evaluated by the untrained, unaided eye. (Two sets of four 7"x10" films, specify positive or negative when ordering.)

GCA T-Ref™
The GCA T-Ref is a laminated paper reference printed with SWOP inks for use by agencies, publishers, separators, printers, and suppliers to determine whether their wide-band reflection densitometer conforms to the ANSI/ISO standard. By assuring conformance to the standard, T-Ref enables users to compare readings from different densitometers with certainty.

To order additional copies of this book or other GCA Publications. . .

Indicate the number of publications/resources you want, complete this form, and return it to GCA. (Prices are GCA member/nonmember. If you are not sure whether you qualify as a GCA member, or if you want to learn more about our quantity and educational institution discounts, please call GCA at 703/841-8190.)

_____	Introduction to Densitometry	$18.45/$34.25
_____	Introduction to Color Bars	$19.25/$39.95
_____	SPECTRUM Guide to Proofs and Pre-Proofs	$10.50/$14.75
_____	GCA/GATF Proof Comparator II (☐ Positive ☐ Negative)	$65.00/$115.00
_____	GCA T-Ref™	$62.50/$85.00

Name/Title _____

Company _____

Address _____

City, State, ZIP _____

Telephone _____

Check Appropriate Boxes:
☐ GCA Member ☐ Nonmember
☐ Check enclosed
☐ Credit Card: ___Visa ___MC ___AmEx

Card Number:_____

Expiration Date: _____

Signature: _____

Please return completed form—include payment and $7.00 ($12 Canada, $20 foreign) for shipping/handling—to GCA, 100 Daingerfield Rd., Alexandria VA 22314-2888. For faster turn-around, call 703/519-8157 to place your order (have credit card information handy). You can also use our fax number: 703/548-2867.

Heterick Memorial Library
Ohio Northern University

DUE	RETURNED	DUE	RETURNED
1.		13.	
2.		14.	
3.		15.	
4.		16.	
5.		17.	
6.		18.	
7.		19.	
8.		20.	
9.		21.	
10.		22.	
11.		23.	
12.		24.